Christian Grollé

COME BACK HOME

THROUGH BACKWARD WALKING

Translated by Linda

Books on Demand

To Rachel, Geneviève, Dominique, Laurence, Edith and Carole.

These wonderful women who have allowed this book to be born.

The advice appearing in this book were lived and verified by the Author. However, neither he nor the Publisher could be considered as liable for any direct or indirect damages due to persons practicing retro movements with excessive or careless behaviours.

ISBN : 9782810616091

SUMMARY

INTRODUCTION

"Come back home through backward walking", why this subtitle ?

Come back home means to trace your steps back to the start.

Backward walk is to step back.

What is the connexion between the two ?

More and more of us suffer in our every day life from the effects of the different crises striking the world, economic, climatic, sociological, medical crises that drive us to depression, degenerative diseases, suicide. A world that runs eyes closed towards, among other things, this statistical aberration promised to us for 2015 : a billion starving people living along two billion obese or overweight people on an earth that could feed all its inhabitants. Can we reasonably go on believing in a humanitarian, responsible and transparent action of those who govern us? What then can be done individually to extricate ourselves from a global crisis? This booklet whishes to convey concret propositions to solve this major questionning through the practise of backward walk.

To begin with, it will allow you "to trace your steps back to the start",

because

To come back home is to step back ; it is the hope of embarked soldiers in wars they do not understand , implemented by financial strategic elites.

To come back home is never to be trapped again by a society that tells us a bigger car, a bigger house, and a bigger salary will bring us happiness.

"Education throughout in the world has failed, it has produced mounting destruction and misery. We have to discover the beneficient ways of creating a new environment; for environment can make the child a brute, an unfeeling specialist, or help him to become a sensitive, intelligent human being. Another and still greater disaster is approaching dangerously close, and most of us are doing nothing whatever about it. We go on day after day exactly as before; we do not want to strip away all our false values and begin anew. We want to do patchwork reform, which only leads to problems of still further reform. But the building is crumbling, the walls are giving way, and fire is destroying it. We must leave the building and start on new ground, with different foundations, different values."

Krishnamurti - *Education and The Significance of Life.*

WHAT IF THE NEW GROUND, THE NEW FOUNDATIONS AND THE DIFFERENT SET OF VALUES WERE...

BACKWARD WALKING ALL THE WAY?

To come back home is to step back from a world that drives us straight into the destruction of human feeling, by enjoining us the faster, higher, stronger, richer, more powerful, more manipulation, more lies. It's high time to understand the power and impact of our individual actions on everyones development.

To come back is to reverse the steps of a civilization that can't satisfy us any longer. Concretely, the big rush forward towards the profit of a few well-to-do to the prejudice of billions of people can be slowed down by practising backward walks. These walks described in this book are as simple as beneficial. Are there, in the world of sports or alternative practises, that many methods having an effect on physical, psychological, energetical and spiritual levels ? As for us, we know of none as complete, and moreover accessible to all….

In backward walking, you will find what you are looking for, depending on your personal concerns of the moment:

- one more training to enhance your physical achievements;
- a medically documented support for a better re-education after a lesion or a wound;
- getting in tune with superior vibratory levels which will concretely magnetize new solutions, meetings, occurrences in your everyday life as well as empower your energetic bodies (and so preparing for the global spiritual change of 2012…)

So, through backward walking, you will be able to "trace your steps back".

Retraction/expansion, flow/ebb, introspection/expansion, backward/forward… Doesn't a popular wisdom say: "step back in order to have a better take-off"? You too can take place in the dances of alternating movements nature bursts with, from the cell to planetary systems.

Backward walking will seem a first vital step in order to take a new start in the world that awaits you. But this time, you will have a new energy liberated through backward walking, an energy that will enable you to fulfil the dreams you always had.

Christian Grollé

www.backward-running-backward.com

BACKWARD WALKING
ALL THE WAY

A VERY ANCIENT HISTORY

Ancestral backward running/walking training in China by the Purple Clouds Temple (Zixiaogong) monks in the Wudang mountains, cradle of Taoist culture.

THE OPPOSITE OF FORWARD IS BACKWARD...

"Alternation is a simple and evident fundametal mechanism. It is relatively easy to understand it and grasp its importance because it is all around us in nature, within us and in the fundamental workings of our bodies: inspiration/expiration, cardiac contraction/expansion, the alternation of day and night, etc. This movement is the fundamental motor of the entire universe and our own biological life. It should be an integral part of our everyday activities and basic habits and serve as a constant reference and way of life. The principle to remember is simply to do the opposite of whatever you are doing, profoundly and systematically, at a deeper and deeper level, at a more conscious, more evolved level."

Dr. Jacques Pezé
The Art of Joyful Living

The principal characteristic of the Tao is the cyclic nature of its ceaseless motion and change. Returning is the motion of the Tao, says Lao Tzu, and going far means returning. The idea is that all developments in nature, those in the physical world as well as those of human situations, show cyclic patterns of coming and going, of expansion and contraction. This idea was no doubt deduced from the movements of the sun and moon and from the change of the seasons, but it was then also taken as a rule of life. The Chinese believe that whenever a situation develops to its extreme, it is bound to turn around and become its opposite. This basic belief has given them courage and perseverance in times of distress and has made them cautions and modest in times of success. It has led to the doctrine of the golden mean.

Fritjof Capra *The Tao of Physics*

THE OPPOSITE OF BACKWARD IS FORWARD...

PRACTICAL ADVICE

First, when walking backwards, you must be careful. As you don't see where you go, you must find a quiet and safe place. Keep away from crowded places (with cars, bikes, children, telephoning pedestrians, animals…) when practicing and, if you have no one to go with you, walk forwards first on the any chosen path to locate any hazards.

If you want to extend your training to backward running*, the best place would be an athletic stadium lane you will have tested running forwards to check its safety.

You can also train with a partner, face to face, one walking forwards the other walking backwards, or side by side alternating forward walking and backward walking together. Practicing this way offers conviviality and some very interesting energetic transfers on more subtle levels*.

But you can also walk at home: a corridor a few meters long is enough, as long as the floor is clear. Make sure you are alone to practice safely. Always start on one foot, take a few steps backwards. Then join your feet, step forward starting on the same foot, taking the same number of steps. Join your feet again, and starting on the other foot, walk backwards and forwards. Repeat these two cycles for about ten minutes, a lapse of time you will modulate according to your training needs.

If you live near the sea shore, you might like to walk/run in rhythm with the flow/ebb of the ocean, without wetting your feet; you will find it is not that easy to merge with the great alternating universal rhythms.

You can also walk backwards along the beach for as long as you wish and come back walking forwards the same length. This exercise can of course be done alone or with a partner, face to face or side by side, but also in a group which is fun.

*see www.backward-running-backward.com, page "Alternative Mixed Running"

It is important to find your place of power, which is a place where you will feel comfortable, peaceful and without any obstacle, far from any critical glances. For my part, I train on a small deserted road, along a river, in the middle of no where, so that, apart from the occasional tractor, I can train safely and peacefully.

I wish you make some wonderful discoveries while backward walking.

BACKWARD WALKING ALL THE WAY

Real progress in understanding nature is rarely incremental. All important advances are sudden intuitions, new principles, new ways of seeing. We have not fully recognized this process of leaping ahead, however, in part because textbooks tend to tame revolutions, whether cultural or scientific. They describe the advances as if they had been logical in their day, not at all shocking. In retrospect, because the bridge of explanation was laid out painstakingly in the years after the intuitive leap, the big ideas seem reasonable, even inevitable. We take them for granted, but at first they sounded crazy.

Marilyn Ferguson - *The Aquarian Conspiracy:*
Personal and Social Transformation in Our Time

FORWARD WALKING

BACKWARD WALKING

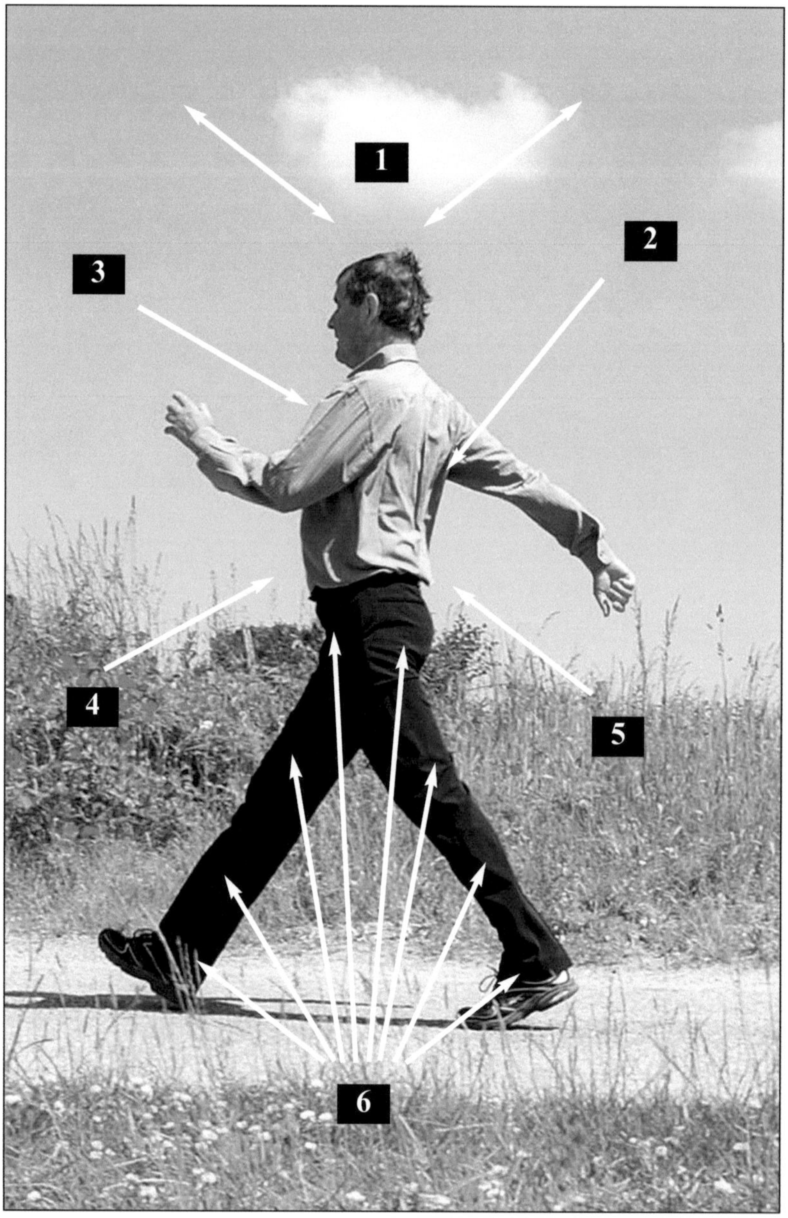

100 STEPS BACKWARDS = 1000 STEPS FORWARDS!

PHYSIOLOGICAL BENEFITS

- Forward walking stimulates the left brain, backward walking stimulates the right brain.
- Alternating synchronizes both brains, bringing about an expansion of consciousness [1].
- Backward walking develops your peripheral and global vision.
- It tends to straighten your spine [2].
- It reinforces your muscles and bones.
- It opens up your rib cage and allows for deeper breathing.
- At equal speed, VO2 (the oxygen volume used) is 68% higher [3].
- At equal speed, HR (heart rate) is 47% higher.
  (156 heartbeats while backward walking against 106 while forward walking).
- At equal speed, VE (ventilation minutes) is 112% higher,
  (these numbers show better cardio-vascular functions).
- The centre of gravity is higher, closer to the sky [4] (spirituality)
  whilst in forward walking it is lower, closer to earth (materiality).
- Backward walking prevents and reduces the risk of impact wounds.
- It improves muscular balance and strengthens your back muscles [5].
- It burns up 25% more calories than forward walking.
- It fortifies and massages your body's internal organs.
- It intensifies your sensory and supra-sensory functions.
- In backward walking, the soles of your feet stand differently*.
- It develops a global coordination which opens you up to Life [6]

* On this subject, see the book Godo, by Dr. Peter Greb.

FORWARD WALKING

BACKWARD WALKING

"What is very interesting and you are never told, is that your body must lean backwards...Yes exactly: you must feel that you are leaning backwards, as if you wanted to lie down on a mattress of air. This way, your heart will be much less brutally stressed than when doing like everybody else: elbows close to your body and your weight heavy on one foot then the other. Your heart should be well looked after, not uselessly exhausted.

[...] Lastly, your breathing – which accelerates at the same time as your heart-beats (and vice-versa) – must be free, natural. The air which is accompanied by part of the regular flux and reflux of cosmic energy, must not be impaired by the weight of your forward bent rib cage"

Noëlle Perez-Christiaens
"La marche, une thérapeutique millénaire"

WHILE WALKING BACKWARDS, YOUR BACK IS TILTED BACKWARDS AUTOMATICALLY...

- IT TAKES SOME STRAIN OFF YOUR HEART,
- IT MAKES YOUR BREATHING EASIER,
- IT STRAIGHTENS YOUR SPINE,
- AND, BY EXTENSION, IT LIFTS YOUR SPIRITS...

NOW IS YOUR TURN TO TRY!

BACKWARD WALKING,
erect upwards

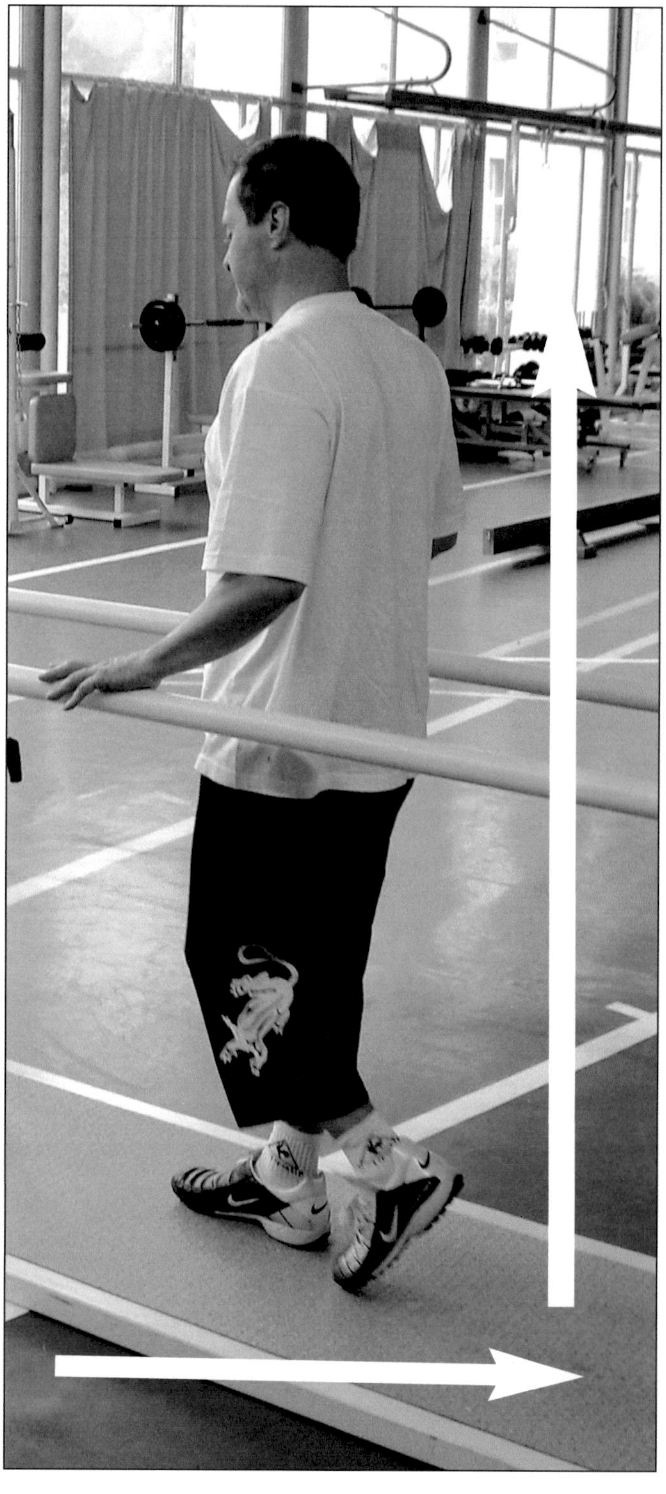

FORWARD WALKING,
bent forwards

THERAPEUTIC BENEFITS

Let our friend Lau Siu On, physiotherapist in Hong-Kong, speak:

"These days, in their professional activity, many people are glued to their chairs for long working hours. The lack of time excuse is not conducive to leaving their chairs to exercise a little. After a few years, a protuberant belly, a lumbar instability and back aches develop. In backward walking, the trunk is tilted backwards slightly, stimulating the abdominal belt and so maintaining the body's balance, which stabilizes the spine and reduces back ache risks. Moreover, backward walking requires the contraction of the gluteous maximus musclesin order to move the legs backwards; this reinforces this muscle which thereby protects our spine.

In forward walking, our centre of gravity is in front of us and so we tend to bend our trunk forwards. In backward walking, it is behind us and we tend to tilt our trunk backwards.

So, backward walking or running can straighten the backs and unfold the shoulders of children with fragile and bent backs. Getting old, this will prevent painful shoulders, neck and back aches.

Backward walking is also recommended for obese or overweight people and who lack time. Their muscles will work more, if their steps are shorter and the rhythm faster: this way 25% of the calories will be burnt up faster than by walking forwards.

A Taïwanese study on 50 older people in need of a re-education program, shows that after a three months' training, the symmetry and the speed of their walk has notably improved.

To conclude: *thanks to our observations, we can testify that backward walking is an excellent re-education and training method, as much to prevent or relieve aches as to improve athletic performances, and above all to preserve one's dearest assets: health and longevity."*

THERAPEUTIC BENEFITS

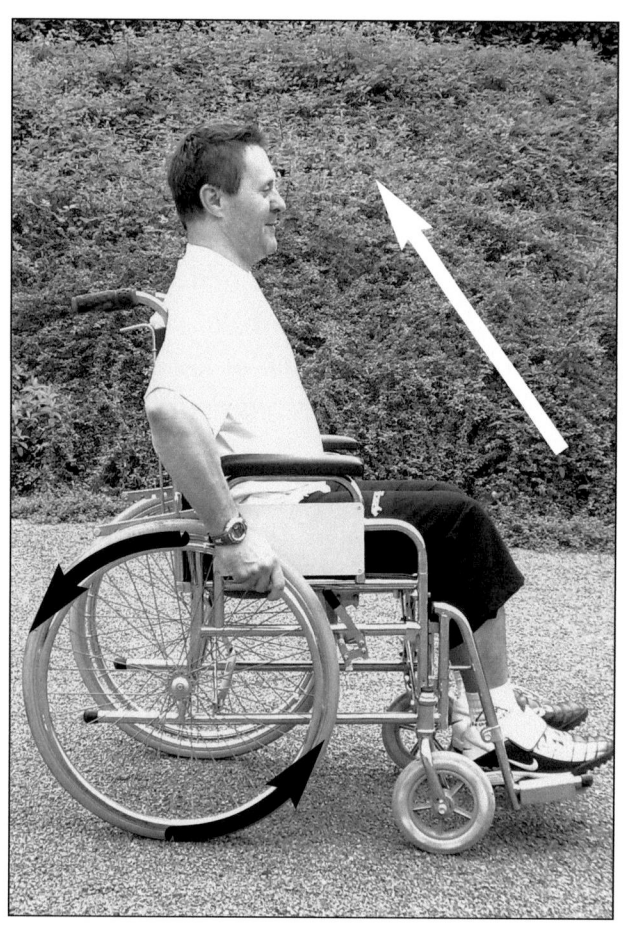

Arms wheel backwards,	**Arms wheel forward,**
body and soul straighten up.	**body and soul coil up.**

Another therapeutic implication of retro movements applies to people in wheelchairs. The idea is to make a new kind of wheelchair with wheels that can be operated, at will, by wheeling backwards or forwards while still driving forwards. The muscular chains and all physiological and psychological systems will straighten up. The spine is straighter, oxygenation is facilitated, spirits get higher as well as the handicapped person's consciousness. Real "miracles" could even be witnessed.

DOCTORS, THERAPISTS, MANUFACTURERS, TAKE NOTICE!

... pedal backwards...

... walk up the stairs backwards...

TO GO BACK IN TIME, AND TO LIVE LONGER, WHY NOT TRY TO...

PSYCHOLOGICAL BENEFITS

OUR SHOULDERS CARRY ALL OUR BURDENS. THE HEAVIER THEY ARE, THE MORE ROUNDED OUR SHOULDERS BECOME.

Watch people walk in the streets around you. All of them are leaning forwards, heads low, eyes looking down, shoulders coiled up. You too maybe… Through a kind of resonance, this closing up entails closed up friendly, love-life and professional relationships, which means they are self-centred. And which trapp people in a terrible solitude. The backpack symbolises the total sum of our burdens. Backward walking will generate a "let go" of this psychological bag so familiar, entailing a straightening of the spine, the head and the eyes, as well as an opening of the shoulders which in turn, will lead to new, more human in the noblest sense, more open, relationships in your life.

The technique is simple. Choose your ground and walk backwards…

BACKWARD WALKING PROMOTES A TRUE "LET GO".

SPIRITUAL LEVEL

MENTAL LEVEL

EMOTIONAL LEVEL

SEXUAL LEVEL

ATHLETIC LEVEL

SOCIO-PROFESSIONAL LEVEL

Backward walking works on all levels of consciousness

RETURNING BACK TO OUR CELL MEMORIES

"The essence of primal therapy is "re-living", which is returning back to a lack of love, early traumas that remained in our system in the shape of imprints. [...] They are stamped in all our systems and our brain as indelible memories that direct our lives. This therapy is a systematic descent into the patient's history, in his past, starting with the more recent problems, and then penetrating in his history through the vehicle of his feelings."

Dr Arthur Janov "Primal healing"

Returning back to our cell memories

All along our life, lack of love, be it parental or other, imprints itself during early childhood, adolescence and adult life, leaving physiological and psychological traces. The essence of backward walking is the "complete re-living"; returning back to our cell memories enables to reverse the process and to start afresh feeling lighter! Walking backwards is reversing the course of things…

Beyond your problems

Symbolically, build a little hillock representing the problems (big or small) you want to get rid of on a starting line. Then walk slowly backwards a hundred meters or more. You will notice that the farther you go, the smaller the hillock gets, until it completely gets clear out of view. Its real proportion has been reached in your new reality… Start forwards again. When you cross the starting line, walk another 100 or 200 meters: The initial problem, as big as it may have seemed, is now part of times gone by.

Magnetizing new solutions

After a backward walk, slow down progressively to a stop. Eyes wide open, observe your feelings, often extraordinary at this precise moment. You have just magnetized new situations, new encounters or events that will bring about unprecedented solutions to what seemed insuperable up to now.

Turn around and walk backwards for 500 meters!

BY CONSTANTLY WANTING TO GO FORTH, ONE REGRESSES COMPLETELY…

BY WALKING BACKWARDS, ONE REALLY ACQUIRES A BROADER OUTLOOK!!!

BENEFITS ON EXCURSIONS

Excursion promoters should include backward walking in their health circuits. If the terrain is appropriate, a sign could suggest you turn around and walk backwards for 500 meters for example, then farther on, another sign could tell you to start walking forwards again, and so on a few times along the circuit. This alternating walks will allow you to recover both physically and psychologically. When getting to an uphill slope, climb it backwards, not only is it less tiring, but you will also see a completely different landscape… Nothing but positive points! The Chinese say:

**BACKWARD WALKING IS EXCELLENT FOR YOUR HEALTH,
LONGEVITY AND SPIRITUAL AWAKENING.**

AGAINST GLOBAL EXTREME POVERTY

SOCIOLOGICAL BENEFITS

"The wise man says it is best to step back in order to have a better take-off.

In its wild race towards progress, our type of civilization – it has been said time and again – needs more wisdom and less knowledge. And wisdom is acquired mostly through indulging in serious reflections on one's self."

Jacques Mayol "Homo Delphinus"

Some years ago, I was in Italy and a backward walk had been organized to make people aware of global warming. More than a hundred of us walked this way. The Italian television ran a report and its impact was quite wide. Imagine thousands of people demonstrating backwards in Paris, New York, Tokyo, Moscow, Peking, New Delhi… against hunger in the world. If we do not understand very soon that we must imperatively start walking backwards on all levels, before long it will become difficult to live on this beautiful Earth.

Backward walking enables you to adopt a more detached perspective, to see the world differently. We all belong to the same Earth, and if we are careless, the cosmos will take its revenge as it has already done in the past. If we do not understand in our bodies, our hearts and our consciences that we are the world, that every thought, feeling and action instantaneously has an impact on humanity, our civilization will not change.

As Krishnamurti said, *"we must abandon the edifice, and start afresh on a new land with new foundations and a different set of values".*

WHAT IF THE NEW GROUND, THE NEW FOUNDATIONS AND THE DIFFERENT SET OF VALUES WERE…

BACKWARD WALKING ALL THE WAY?

COSMIC BODY

THE ASSEMBLAGE POINT →

THE ENERGETIC BODY IS WALKING **FORWARDS**

THE PHYSICAL BODY IS WALKING **BACKWARDS**

ENERGETIC BENEFITS

Initiated by a well known Philippine healer, Dr Janine Fontaine learns how to feel the energetic and spiritual bodies of a human being. She keeps healing and passing on her knowledge on the three bodies of the human being (physical, energetic, spiritual or cosmic).

How do these three bodies work while walking backwards?

While the physical body walks backwards, the energetic body walks forwards and vice versa. This double physical/energetic alternating allows for the cosmic body to radiate with a superior dense/dance energy. To discover the movements of our conscience through these alternating walks is a great bliss I hope you will enjoy.

"Don Juan contended that our world, which we believe to be unique and absolute, is only one in a cluster of consecutive worlds, arranged like the layers of an onion. He asserted that even trough we have been energetically conditioned to perceive solely our world, we still have the capability of entering into those other realms, which are as real, unique, absolute, and engulfing as our own world is."

[...] "In the course of his teachings, Don Juan repeatedly discussed and explained what he considered the decisive finding of the sorcerers of antiquity. He called it the crucial feature of human beings as luminous balls: a round spot of intense brilliance, the size of a tennis ball, permanently lodged inside the luminous ball, flush with its surface, about two feet back from the crest of a person's right shoulder blade. He said that the old sorcerers named it the assemblage point after seeing what it does. "What does the assemblage point do?" I asked. "It makes us perceive," he replied. "The old sorcerers saw that, in human beings, perception is assembled there, on that point."

Carlos Castaneda *"The Art of Dreaming"*

BACKWARD WALKING MAKES IT POSSIBLE TO BE CONNECTED WITH THE ASSEMBLAGE POINT AND WITH THE COSMIC BODY. THIS TECHNIQUE IS OPEN TO ALL! TRY IT!

FOR PREGNANT WOMEN TOO... AND CHILDREN
LET'S GET INTO REVERSE
TO GIVE LIFE A CHANCE!

FORWARD
WALKING

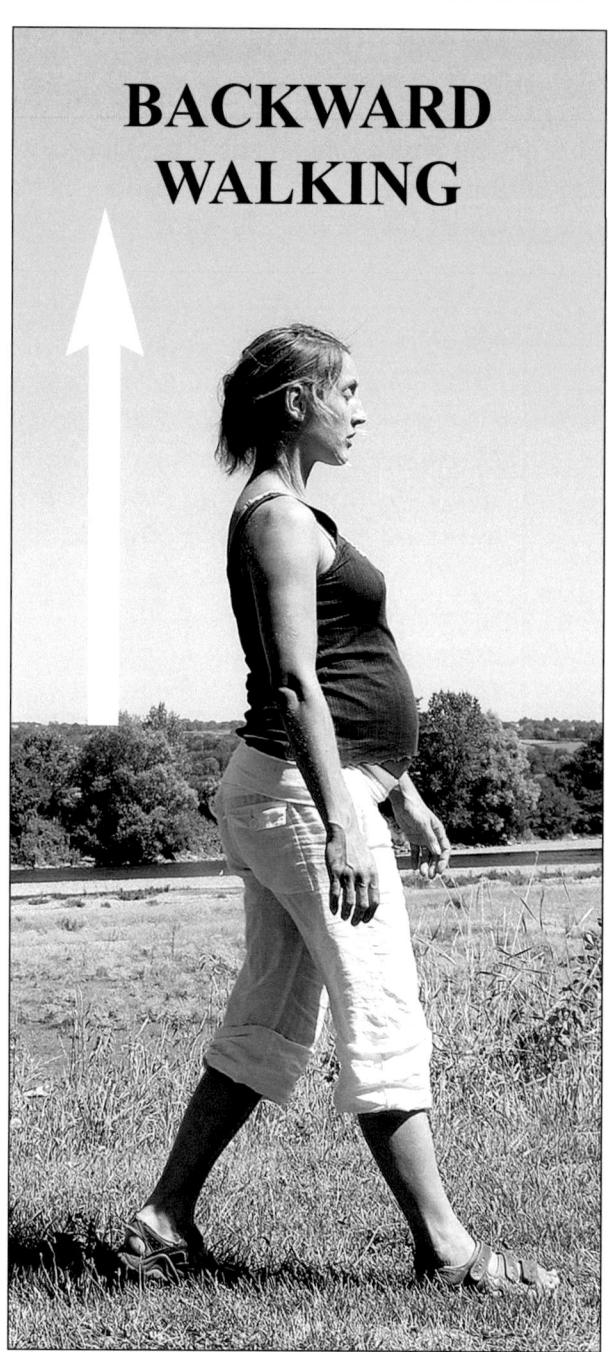

BACKWARD
WALKING

COME BACK HOME
THROUGH BACKWARD WALKING

All of us aspire to trace our steps back to home. The Zen awakening masters talk about Satori which means "Returning home". Returning to the cosmos where we come from. A thousand year old Wisdom says there is a time to explore the world: that is the journey there (forward walking), and a time to get back home: that is the journey back (backward walking). Both make up the Tao. Salmons swim downriver to explore the vast ocean then swim upstream to beget life and die. Legend has it that some make this journey twice and become immortal... Backward walking allows to become younger again and to step back to the origin of all things:

OUR TRUE "HOME"

Every day, I walk 2000 meters: 1200 meters backwards and 800 meters forwards.

The 60% backwards and 40% forwards ratio seems excellent because one needs to do more backward walking in order to reduce global poverty.

GOOD EXPERIMENTING!

CONCLUSION

Adapt your training to your personal capacities, 10 minutes backwards a day in your corridor, 20 minutes during your usual excursions… The main thing is to get started and to keep on going confidently and according to your capabilities.

The more you walk backwards, the more you will want to. As if nature claimed its due after thousands of years of excessive forward flight. Perhaps are we going to rediscover the path to wisdom and, paradoxically, to common sense.

Much could still be said on the benefits of backward walking, but the main thing is not the description of it, it is the experimenting in your very own body. The map is not the territory and as long as you have not inwardly felt the innumerable profits of this new – and ancestral! – way to walk (or run), you will know only one half of the history of the universe; you will know only one half of yourself.

The time for folly comes to an end, here comes the time for global harmony.

To conclude, this is what Tchouang-Tseu says:

"Life is a harmonious mixture of Yin and Yang"

SCIENTIFIC ANNEXES

"Retro ambulation, whether it be backward walking or backward running, is an extremely effective form of... rehabilitation. A comprehensive understanding of the biomechanics of retro allows this form of rehabilitation to be appropriately used for certain conditions affecting the foot and ankle, lower leg, knee, thigh, hip and pelvic and low back regions. Because of the unique forces, stresses and motions facilitated by walking or running backwards, retro, used with clinical control, is becoming a primary and full spectrum form of rehabilitation. It is also used quite extensively for cross training and sports-specific conditioning."

[...] "Backward walking or running is clearly not a mirror image of forward walking or running. Human backward walking is not achieved by a simple change in phase relationships between the hip and knee joints. The marked changes in the movements of the joints and their interactions reflect different demands placed on the system during backward walking. These differences can be used to better understand the interaction between neuro control mechanisms and functional responses during human locomotion."

Gary Gray, P.T. *"Retro-Rehab... an Extremely Effective Form of Rehabilitation &Training ! "*

http://www.backward-running-backward.com/PDFGaryGray.pdf

"Backward walking and/or running can potentially provide unique benefits to the individual rehabilitating an injury as well as to the exercise enthusiast who is facing the inevitable effects of age and past specific training on the body. Some conditions that have been observed to respond positively to retro locomotion include the following: muscle strains including low back, hip, groin and hamstrings; ankle sprains; post-surgical knee joint rehabilitation; shin splint syndrome; achilles tendon strains."

[...] "This alone can be a benefit if one is suffering from an impact-type injury which could manifest as a sore knee, stress fracture, or similar problem or has not been recently engaged in an active exercise program. [...] A treadmill with rails can be used to assist greatly in a forward and/or backward walking rehabilitation program. Since a lower extremity injury often causes a reduction in strength and/or pain in the affected limb, partially supporting body weight by using the rails can be helpful."

Professeur Emeritus Barry Bates, Ph.D.,
Président of *Human Performance & Wellness.*
"Forward and backward locomotion, understanding the benefits".

http://www.backward-running-backward.com/UNDERSTANDPDFFRENCH.pdf

JOGGING

RETRO RUNNING

"Surprisingly, people could also walk easily with one leg moving forward and the other backward, a pattern referred to as "hybrid walking." Adaptation of hybrid walking, in which varying speeds were applied to legs walking in opposite directions, was found to interfere with subsequent "normal" forward and backward walking. The combined results demonstrate there are distinct brain modules responsible for right/forward, right/backward, left/forward and left/backward walking. Most significantly, these modules can be individually trained, which would be critical for rehabilitation focused on correcting walking asymmetries produced by brain damage. The notion that we can leverage the brain's adaptive capacity and effectively 'dial in' the patterns of movement that we want patients to learn is incredibly exciting," [...] "These findings significantly enhance our understanding of motor skills, effective therapeutic approaches and the true adaptive nature of the brain."

Dr. Amy Bastian, *Director of Analysis Movement*
Laboratory at Kennedy Krieger Institute (Baltimore)
"New research discovers independent brain networks control human walking".

htpp://multivu.prnewswire.com/mnr/bastian/29010/

RETRO RUNNING

"Forward and backward body locomotion constitutes a pure form of approach and avoidance behavior with high ecological validity. Corroborating previous evidence that avoidance cues facilitate the recruitment of cognitive control, the current study showed that stepping backward significantly enhanced cognitive performance compared to stepping forward or sideways. Considering the effect size, backward locomotion appears to be a very powerful trigger to mobilize cognitive resources. Thus, whenever you encounter a difficult situation, stepping backward may boost your capability to deal with it effectively".

Docteur Koch, Severine, R. W., Hengstler, M., & van Knippenberg, A.
"Body locomotion as regulatory process: Stepping backward enhances cognitive control"

http://www.ru.nl/socialpsychology/faculty/dr_severine_koch/

"Theory predicted that we could send light backwards, but nobody knew if the theory would hold up or even if it could be observed in laboratory conditions." [...] "It's weird stuff. We sent a pulse through an optical fiber, and before its peak even entered the fiber, it was exiting the other end. Through experiments we were able to see that the pulse inside the fiber was actually moving backward, linking the input and output pulses. So, wouldn't Einstein shake a finger at all these strange goings-on? After all, this seems to violate Einstein's sacred tenet that nothing can travel faster than the speed of light. So, wouldn't Einstein shake a finger at all these strange goings-on? After all, this seems to violate Einstein's sacred tenet that nothing can travel faster than the speed of light. So, wouldn't Einstein shake a finger at all these strange goings-on? After all, this seems to violate Einstein's sacred tenet that nothing can travel faster than the speed of light."

Professeur Robert Boyd
"Light's Most Exotic Trick Yet : So Fast it Goes ... Backwards?"

http://www.rochester.edu/news/show.php?id=2544

"Backwards walking or running (marcha reversiva) breaks paradigms, concepts, neuromuscular patterns, and alters our world view. When, standing up on a chair, one observe things, one can detect some changes in the environment, new perceptions and details. The same thing happens if we look the world in reverse: we can see many differences, and when we return to the « normal » position, we notice the numerous stimulus that we don't experience in our daily life. To be able to appreciate beauties and feelings the world offers, it's necessary to live and interact with it; In combining the body with the soul and the mind which are nothing more than divisions of one and only structure of life.

Don't be afraid to test things we don't know; let's be curious and respectful of the unknown. Backwards walking or running gets an extraordinary work of physical, emotional, mental and spiritual recovery. With this practice in the respect of your own body, you could feel that the movement becomes light, subtle, balanced and every new stimulation will be as fluid as the river's current, always open to the immensity of Life."

Professor Pablo Galleto (Brazil)
http://www.marchareversiva.com/

Newly Forming Solar System Has Planets Running Backwards

"Call it the biggest beltway ever seen. Astronomers have discovered a newly forming solar system with the inner part orbiting in one direction and the outer part orbiting the other way." [...] "Our solar system is a one-way boulevard. All the planets --- from Mercury out to Pluto and even the newly discovered objects beyond --- revolve around the Sun in the same direction. This is because the Sun and planets formed from the same massive, rotating cloud of dust and gas. The motion of that cloud set the motion of the planets.The fact that a solar system can have planets running in opposite directions is a shocker. This is the first time anyone has seen anything like this, and it means that the process of forming planets from such disks is more complex than we previously expected," said Anthony Remijan of the National Radio Astronomy Observatory. This is the first time such a phenomenon has been seen in a disk around a young star. Yet who's to say the arrangement is uncommon? As astronomers find more and more extra-solar planets (over a hundred so far and counting), they are realizing that solar systems come in many shapes and sizes."

NASA : "Newly Forming Solar System Has Planets Running Backwards".

http://www.nasa.gov/centers/goddard/news/topstory/2006/opposite_orbit.html

For more information on the differences of VO2 between forward and backward, see: Flynn T.W., Connery S.M., Smutok M.A., Zeballos R.J., Weisman I.M *"Comparison of cardiopulmonary responses during forward and backward walking and running in normals"*. Department of Clinical Investigation, William Beaumont Army Medical Center, El Paso, TX 79920-5001.

http://www.ncbi.nlm.gov/pubmed

MIXED RUNNING

BACKWARD RUNNING
ALL THE WAY

"The emergence of backwards walking and running is not a matter of chance, it is both a sign and the symbol that something "in the air" is becoming reversed. This new way of walking announces the beginning of a new civilization. To reverse the most basic and common act of man, walking or running, is tantamount to touching his essence. What could be more effective than this simple exercise which will modify our view of everything? It can change our human relationships, our relationship to money, our conception of time, and at the social level can alter the face of our cities, our transportation and exchange systems, etc. At a time when an unprecedented human mutation is taking place before our very eyes, if practided as an exercise conducive to a change in our perceptions, the advent of backward running/walking can bring down the certainties that underlie the general escapist movement. More than words can, a backward motion of our bodies will provoke a fundamental questioning of our commonest attitudes and bring about the social change that the urgency and gravity of the problems of our day require."

Professor Patrick Baronnet
Retro Running, une autre manière de courir

BIBLIOGRAPHY

Carlos Castaneda, *"The Art of Dreaming" Elements Books, 2004.*

Christian Grollé, "*Retro Running, une autre manière de courir*", Auto-Édition 1994.

Dr. Arthur Janov *"Primal Healing: Access the Incredible Power of Feelings to Improve Your Health", New Page Books, 2006.*

Dr. Jacques Pezé, *"The Art of Joyful Living" Element Books, 1991.*

Dr. Janine Fontaine , *"La Médecine du Corps Énergétique",* Robert Laffont, 1983.

Dr. Peter Greb, "*Godo, mit dem Herzen gehen*", Koha-Verlag 2000.

Fritjof Capra, *"The Tao of Physics: An Exploration of the Parallels between Modern Physic and Eastern Mysticism" Shambhala Publications, 2000.*

Jacques Mayol, *"Homo delphinus, The Dolphin Within Man" Idelson-Gnocchi, 2000.*

Krishnamurti, *"Education and The Significance of Life" HarperOne, 1 Edition 1981.*

Marilyn Ferguson - *The Aquarian Conspiracy: Personal and Social Transformation in Our Time,* J.P. Tarcher 1987. - *Aquarius Now: Radical Common Sense And Reclaiming Our Personal Sovereignty,* Weiser Books 2005.

Noëlle Perez-Christiaens, *"La Marche, une Thérapeutique Millénaire",* B.K.S. 1980.

Sri Aurobindo *"Collected Poems",* Lotus Press, 1972.

WEBSITES

Italian Retro Running Association Giuseppe Angeli (Italy) http://www.retrorunning.com

International Retro Running Roland Wegner (Germany) http://www.retrorunning.de

Dr. Robert K. Stevenson (United States) http://www.backwardsrunning.com

Professor Pablo Galleto (Brazil) http://www.marchareversiva.com

Lau Siu On (Hong Kong) http://www.lausiuon.com/BW.htm

Johannes Gosch (Austria) http://www.timelessvision.at/

Professors Barry Bates & Janet Dufek (United States)
http://darkwing.uoregon.edu/~btbates/index.html

All is not finished in the Unseen's decree !
A mind beyond our min demands our ken ;
A life of unimagined harmony
Awaits, concealed, the grasp of unborn men.

The crude beginnings of the lifeless earth
And mindless stirrings of the plant and tree
Prepared our thought ; thought for a godlike birth
Broadens the mould of our mortality.

A might no human will or force could gain,
A knowledge seated in eternity,
A joy beyond our struggle and our pain
Is this earth-hampered creature's destiny.

O Thou who climbed to mind from the dull stone,
Turn to the miracle summits yet unwon.

Sri Aurobindo - Collected Poems.

Christian Grollé

c.grolle@orange.fr

http://www.backward-running-backward.com/

Books on Demand GmbH,

12/14 Rond-Point des Champs-Elysées, 75008 Paris, France.

Printed by Books on Demand GmbH, Norderstedt, Germany.

April 2010